CONTENTS

This Place Is Special 1
What's Up? 4
Why the Mess? 6
How You Can Change Things 7
   Your Goals 7
   Your Scale of Action 7
   Your Direction of Action 8
   How Much Is Enough? 9

50 SIMPLE THINGS YOU CAN DO TO SAVE HAWAI'I . . . 10

Preserve the pristine places 12
   1. Monitor and protect wildlife 12
   2. Restore your favorite ecosystems 13
   3. Create a pristine Hawaiian place 14
   4. Don't support alien species invasions 15
   5. Reduce the number of feral animals 16
   6. Don't support poaching 17
   7. Know where things you buy come from 19
   8. Calculate your ecological footprint and shrink it 20
   9. Reproduce responsibly 21
   10. Support conservation-oriented policy changes 22

Keep coral reefs healthy 23
   11. Don't contribute to polluted runoff 23
   12. Prevent marine debris 25
   13. Protect our shorelines 27

Use the ocean wisely 28
   14. Fish responsibly 28
   15. Eat seafood responsibly 29

Reduce solid waste 30
   16. Live simply 30
   17. Avoid overpackaged and disposable items 31
   18. Keep your drinking water clean and simple 32
   19. Go paperless 33
   20. Reuse 34
   21. Repair 35
   22. Recycle bottles 36
   23. Recycle aluminum, plastics, white paper, newspaper, paper bags, cardboard, and glass 37
   24. Recycle the odd stuff 38
   25. Give away or sell unwanted items 40

Keep your home hazardous-waste free 41
   26. Don't generate hazardous waste 41

Save water 42
   27. Use less water inside 42
   28. Use less water outside 44

Conserve energy 46
   29. Burn calories, not fossil fuels 46
   30. Use lighting efficiently 47
   31. Cool your home efficiently 48
   32. Buy energy-efficient appliances 49
   33. Reduce appliance use 50
   34. Choose renewable energy sources 51
   35. Perform an energy audit 52
   36. Lobby for more and better energy bills 53

Reduce global warming emissions 54
   37. Drive less 54
   38. Use less energy when you drive 55

Prevent pollution 56
   39. Buy organic food 56
   40. Prevent our next sewage spill 57
   41. Support politicians who improve public infrastructure 58

Promote a culture of loving the land 59
   42. Learn about the land 59
   43. Teach about the land 61
   44. Eat whole foods 62
   45. Support Earth-friendly and locally owned establishments 63
   46. Invest in environmentally responsible companies 64
   47. Don't build or buy in high-risk areas 65
   48. Travel responsibly 66
   49. Get involved at work 67
   50. Get involved in the community 68

Appendix: Hawai'i's Ten Major Environmental Issues 69
   Pristine Places 69
   Coral Reefs 71
   Ocean Resources 74
   Solid Waste 76
   Hazardous Waste 80
   Water Consumption 82
   Energy 85
   Global Warming 88
   Pollution 89
   Culture 92

Sources 93
Directory of Environmental Organizations and Agencies 95

THIS PLACE IS SPECIAL

The *Kumulipo,* a Hawaiian origin chant, begins with the coral polyp and ends, many, many named Hawaiian creatures later, not with humans, who are recognized to be sure, but with *kalo* (taro), on which the Hawaiians depended and therefore treated with respect, like an elder. Such is the tradition those of us who were born in Hawaiʻi or come to Hawaiʻi inherit. It is a strong tradition of knowing, respecting, and caring for our place.

Early Hawaiians had a deep connection to the land. To them, the *ʻāina* (land) was sacred. Many Hawaiians today share this connection to and gratitude for what the land provides: food and materials, spiritual health, and a home for themselves and other living creatures.

Those of us who are not native Hawaiians may not have the same feelings Hawaiians do for the land, but we appreciate its extreme natural beauty and are thankful to be in the presence of that beauty whether for just a few weeks on a visit or for a lifetime.

When we appreciate a place, we are more likely to care for the place. As Michael Soulé, founder of the Society for Conservation Biology, said, "Facts about extinction compute, but they don't often convert."

But some people like facts. And many facts bear out what a special place Hawaiʻi is. Unfortunately, although many individuals and organizations make it their business to know the facts, public awareness has not reached the "tipping point." The Hawaiʻi Visitors and Convention Bureau's Web site image gallery has only two pictures of Hawaiian animals, a *nēnē* (Hawaiian goose) and a *honu* (green sea turtle). While both are endangered species, unique to the Hawaiian Islands, the vast majority of Hawaiʻi's 23,149 additional species are unique as well, and hundreds of them are currently listed as endangered by the U.S. Fish and Wildlife Service.

Hawaiʻi is the remotest archipelago in the world. The eight main islands, Hawaiʻi (the Big Island), Maui, Kahoʻolawe, Lānaʻi, Molokaʻi, Oʻahu, Kauaʻi, and Niʻihau, compose just the southeastern, 600-kilometer-long end of the Hawaiian chain. The chain extends another 2,000 kilometers northwest of Niʻihau as the Northwestern Hawaiian

Islands. The youngest emergent island in the chain is the southeasternmost Big Island; the oldest is the northwestern-most Kure Atoll. Kure is 60 million years older than the Big Island. The northwestern islands are much lower and smaller because they have sunk and eroded over the millennia. Nevertheless, they are treasures to native Hawaiians for their historical and cultural significance and to people worldwide for the near-pristine and protected coral reefs that hug their remaining shallow reaches.

Because of Hawai'i's remoteness, any terrestrial creature that has arrived in Hawai'i without human help either got here by surviving a very, very long journey or evolved here from those lucky arrivals that managed to survive and reproduce. Hawai'i's creatures are therefore special both as a unique assemblage, found nowhere else on Earth, and as individuals, since over 90% of Hawai'i's native land and fresh water species exist only in Hawai'i. In some cases, species are found in only one place in the islands—on one mountain range, for example.

Hawai'i's ocean ecosystems are also notable. Hawai'i is so remote that 25% of its marine species are unique. And just like Hawai'i's terrestrial ecosystems, its marine ecosystems contain unique assemblages of organisms.

Hawai'i is also geologically impressive. The volcanic slopes of Mauna Kea and Mauna Loa, which rise over 13,000 feet above sea level, are the tallest mountains in the world if measured from the sea floor. Hawai'i also boasts the tallest sea cliffs in the world and the longest continuous volcanic eruptions. Certain places in Hawai'i, the top of Mt. Wai'ale'ale, on Kaua'i, for example, receive some of the highest rates of rainfall in the world, an average of 10 meters (400 inches) per annum. Only a few miles away, as an 'alalā (Hawaiian crow, now nearly extinct) flies, are places that receive fewer than 20 inches a year.

> **native, or endemic:** originating in a particular place and found only in that place.
>
> **indigenous:** native to a place but found in other places as well.
>
> **exotic, or introduced:** brought to a place rather than originating there.
>
> **tipping point:** the point at which we begin to collectively believe something or take action.

When altitude and time combine with moist trade winds, rain-driven erosion changes the face of islands. Hawai'i's islands are in all stages of the aging process. The islands also have more types of microhabitats, over 150, than anywhere else on Earth. So even though the land area of the islands is small, it includes a huge array of diverse places for creatures to live.

Hawaiian culture, like Hawaiian creatures and geology, is special. The people who found and thrived on these remote islands over 1,500 years ago had courage and skill. Over a long period, they acquired a collective wisdom about this place, its natural history, and its value. Today, much less of this knowledge is instilled in Hawai'i's diverse human inhabitants, but it does still exist and is being revived and shared more and more.

50 Simple Things You Can Do to Save Hawai'i aims to inspire you to *mālama* (care for) the *'āina*. With this book as a guide, we can keep Hawai'i special for our grandchildren *and* the descendants of the coral polyp.

WHAT'S UP?

All the special things about Hawai'i, whether you consider them emotionally or rationally, are many and diverse, and they apply to a very small place. It has often been said that we shouldn't design our cities, dispose of our trash, build roads, or use energy as if we are part of a continent. We need uniquely Hawaiian island dreams and success strategies.

One of the good things about Hawai'i's size (0.29% of the total U.S. area) is that we know, better than people in places with a large area, exactly what's going on, or "wassup?" as we like to say. We see or hear what happens directly, or we hear from friends, or friends of friends who have friends who know what's up. This means that we have a chance to "manage the store" better than people from larger places.

From some people's perspectives, all is well in Hawai'i. We have a strong economy with a high employment rate, and the solid beginnings of a revival of Hawaiian culture. Green sea turtles are returning to the reefs after being protected since 1978, and the trade winds blow most of the time. Nevertheless, there *is* some trouble in paradise, and it won't go away unless we find it out, face it, and fix it.

Those of us who care about the health of the environment often become discouraged: by sewage gushing into the Ala Wai, mud covering a coral reef in Kaua'i, pollution-emitting traffic jams in Kona on the Big Island, or the water shortage in Maui a few years back. There are a number of ways, however, that we can take feelings of sadness, anger, dismay, or helplessness and use them to produce positive outcomes.

First of all, those of us who want to keep Hawai'i's environment healthy should be proud that we notice that something is up. We cannot help solve a problem unless we acknowledge that there is one. (Researchers have hypothesized that those who deny that environmental problems exist do so because those problems occur gradually or are complicated and difficult to fix.)

Even those who don't acknowledge problems may help fix them, however. For example, many people recycle bottles now because of our beverage container deposit program. In so doing they help alleviate our solid waste disposal problem. Some of them don't recycle for that reason, however. They do it for the cash.

Fortunately, our individual actions have larger support. The ten sections in the Appendix describe Hawaiʻi's ten major environmental issues, and there isn't one of them that the federal, state, or city and county government or the private sector is not doing something about. For every issue, many ongoing or planned actions aim to ameliorate the problem. We are not a complacent place. Hawaiʻi's Integrated Solid Waste Management Plan says that the state will "work to maximize all elements of the 'triple bottom line' (i.e., economic prosperity, environmental stewardship and social equity), and not to trade one against the other." Hawaiʻi's new Sustainability Task Force is gearing up to revise the Hawaiʻi State Plan and create a Hawaiʻi 2050 Sustainability Plan by 2008. The goals of the Plan will be to achieve sustainable communities, a sustainable economy, a sustainable quality of individual and family life, and a sustainable environment. Sustain Hawaiʻi, a nonprofit organization, defines sustainability as "The capacity to provide the best of ourselves, each other and all things in our environment now and in the future."

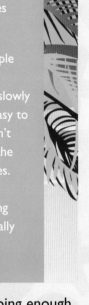

sustainable growth: This an oxymoron. Growth of any kind that uses up nonrenewable resources cannot be sustained.

sliding/shifting baseline: People tend to perceive ecosystems as healthy even though they have slowly and steadily deteriorated. It's easy to do because gradual declines don't register on the disaster meter the same way a massive oil spill does. Yet gradual declines have taken ecosystems to the same alarming endpoint that is more dramatically and obviously wrought by large destructive events.

Are people in Hawaiʻi doing enough to address the challenges we face? Some people think we are simply making ourselves feel good through small efforts, while others think we are making real, effective, and sufficient changes. The point of this book is not to resolve this question. This book is for the doers. And for the doubters, those who think we need to do more. And for those who have never done a thing and want to start now! The only problem that is unique to Hawaiʻi is the loss of Hawaiian cultural values; all the other challenges we face occur everywhere. We have a chance to become trendsetters for the world in our special, smaller place, where anything is possible.

WHY THE MESS?

These five beliefs are most often cited to explain attitudes toward Hawai'i's environmental issues:

Maybe the loss of species, coral reef degradation, filling landfills, and other environmental issues are not really problems.
Example: Traffic: don't judge it, accept it; it's natural.

Many of the problems we face today are the unforeseen negative outcomes of good and innocent efforts. The problems emerged out of decent behaviors that, when multiplied, led to trouble.
Example: Traffic: a simple problem of too many cars and not enough road.

Technology and the values that encourage it have done more harm than good.
Example: Traffic: Replace cars with bicycles; redesign communities.

Global and local population growth causes or exacerbates our environmental problems.
Example: Traffic: cars would be fine if there weren't so many of us driving them.

We need to adopt a more holistic, ecocentric value system.
Example: Traffic: we need to invest money and energy in fixing it instead of tolerating the unhealthy status quo!

The good news about this list is that it suggests solutions: Don't accept what you don't like. Don't blame, because in many ways we are all innocent; almost no one set out to cause Hawai'i's environmental problems. Technology can be a fix, but we need to be wary; no one believes technology is a cure-all anymore. We must be aware of the effects of population growth in this state; it is a huge challenge to the ongoing health of Hawai'i. And finally, we need to have a vision of what Hawai'i ought to be and not sell ourselves short for smaller, individual, ephemeral gain.

HOW YOU CAN CHANGE THINGS

My paddling coach, the late Irwin Keliʻipuleʻole, taught our crew a new word one day when a new canoe was graced the *Laulima*. That's Hawaiian for "pull together." And as with most Hawaiian words, its meaning has many layers. It implies that people *have to* work together in order to be successful. Irwin explained this with a story. He told us that when early Hawaiians fished, the whole village came out. Some people slapped the water, others lay a long net, and others made sure fish didn't swim out the sides as the net was walked to shore. In the end, if all did their parts, there was food for the whole village. In a canoe, as with fishing, *all* must pull together. Pulling together is also the way to keep Hawaiʻi healthy.

The simple way to make Hawaiʻi's environment healthier is through action. In many cases we even know what kinds of actions we need to take: use much more renewable energy, landfill less trash, stop polluted runoff, devise better mass transit, quit overfishing herbivorous reef fish, and so on. We also know that we are doing a number of these things already, and have more in the works.

In other cases, we know what we need to achieve, but the process of initiating collective action is bogged down by details or is limited by the amount of time each of us can dedicate exclusively to helping Hawaiʻi remain healthy. We have school to attend, children to raise, jobs to fulfill, and miles to paddle in our canoes. If you think about it, each of us is somewhat like the state: big dreams, finite resources!

How do you decide where and how to take action? The following section is designed to help you come up with your own personally satisfying recipe for helping Hawaiʻi.

Your Goals

All of us share the same major goal: We want to make the greatest difference for Hawaiʻi's future and enjoy the process. Ask yourself, What can I do that 1) *I would enjoy*, and that 2) *I know I can do*, and that would 3) *make the greatest difference* in (4) *an efficient way?*

Your Scale of Action

We all act on many levels every day. Sometimes we go it alone; other times we work together with family, friends,

action—many little efforts that don't require a grand plan. The treaty that led to the global banning of the chemicals that caused the ozone "hole" is an example of the second kind of action—an agreement that took time, research, and money to develop and implement.

The 50 Simple Things presented for you to act on involve many levels of actions. Do the things that reap the greatest rewards for you and Hawai'i.

Your Direction of Action

Are you connected enough to the "movers and shakers" in a particular arena to go directly to them and make changes there from the top down? If you go to the legislature with an idea for a bill that will save the state energy and money and you get some legislators to write and sponsor "your" bill, you have worked from the top down. If you go to your boss with an idea for involving the entire company in a service project that will bring goodwill to the company and make the forests healthier, and your boss champions the idea and develops an incentive plan for employees, you have again worked from the top down.

coworkers, community, island, state, nation, or even the world. Remember when Hawai'i's Little League team went to the world championships? When Jasmine Trias made it to the final round of *American Idol*? We definitely know how to come together as a state.

Sometimes we reach environmental goals when many small individual actions add up. Other times groups act together in one big effort. Recycling household trash is an example of the first kind of

Or are you more connected to the people, the grassroots, the majority of us? When the students at a high school or university organize their own whitepaper recycling effort on campus, without any direction or inspiration from administrators, it's a grassroots effort. When the Hawaiian community makes its own community fishing plan to prevent overfishing, it's a grassroots effort.

Or perhaps you make a plan and engage your peers and work as a tribe of equals. An example is a team of paddlers who host a canoe race in which a percentage of the proceeds goes to a charitable organization that dedicates itself to coral reef protection.

All of these strategies work. Consider the issues, the outcomes, and your preferences when you decide to act.

How Much Is Enough?

Doing nothing is not enough, and dedicating ourselves completely to the environment and nothing else is more than we can all realistically hope for. The book *Simple Things Won't Save the Earth* makes the point that if we all do just a few simple things then we won't "save" the planet; it won't be enough, and our own and future generations will live in a world that is less and less healthy, beautiful, and capable of providing for us and other species into the future.

How much of each Hawai'i-saving activity is enough? We do know that doing something is much better than doing nothing. So do what you can, when you can, and inspire others to do so as well. Whenever you can, add an activity to the list of what you do. Whenever you can, capacity-build your efforts: make it easier, more enjoyable, more rewarding, more attractive, more morally correct for people to take care of Hawai'i.

50 SIMPLE THINGS YOU CAN DO TO SAVE HAWAI'I

The following actions don't require special training, a certificate, a degree, or approval. They are particularly beneficial to Hawai'i, targeting each of Hawai'i's ten major environmental challenges: the loss of pristine places, coral reef degradation, overharvesting of marine resources, solid waste management, hazardous waste management, water consumption, energy, global warming, pollution, and the loss of culture. They use the latest technologies and apply new ideas to achieve very old, very noble goals that are at the core of traditional Hawaiian values. Doing these things will nurture, restore, protect, and show respect for the 'āina.

When you're recycling your cans or waiting to testify in a crowded room for a bill you believe will help make Hawai'i a healthier place, stop for a moment and feel the joy in making a difference, the strength in acting with others, the harmony in carrying out a Hawaiian tradition in a new way. Don't forget this part. Besides the visible difference you make, this is your reward.

1 Monitor and protect wildlife.

PRESERVE THE PRISTINE PLACES

- **Participate in the annual Audubon Society Christmas bird count.**
  www.hawaiiaudubon.com/xmas.html

- **Participate in a Reef Check survey.**
  www.reefcheckhawaii.org/

- **Join the Blue Water Response Team.**
  www.hi.sierraclub.org/bluewater/

- **Hunt for feral pigs, goats, or deer the Hawaiian way.**
  www.hawaii.gov/dlnr/dcre/know.htm
  www.hawaii.thehuntingtrail.net/

Restore your favorite ecosystems. 2

- **Restore forests through alien plant eradication and native plant outplanting.**
  www.nature.org/wherewework/northamerica/states/hawaii
  www.hi.sierraclub.org/hstp/index.html

- **Turn your greenhouse into a native plant restoration nursery.** For information on growing native Hawaiian plants, read *Growing Native Hawaiian Plants,* by Heidi Bornhorst.

- **Remove alien algae from Hawai'i's reefs.**
  www.hear.org/cgaps/events/alienalgaecleanup/index.html

PRESERVE THE PRISTINE PLACES

PRESERVE THE PRISTINE PLACES

3 Create a pristine Hawaiian place.

Heidi Bornhorst

- Grow native plants in your yard, in pots, and at work. The Hawaiiana section of your local bookstore offers several titles about growing native Hawaiian plants.

  For a list of some of the nurseries in Hawai'i that specialize in selling native Hawaiian plants, visit
  www.hawaii.gov/health/oeqc/garden/eioegsrc.htm

- Educate groundspeople at home and at work about the pluses of using native Hawaiian plants in landscaping. Learn more at
  www.hbws.org/

Heidi Bornhorst

Don't support alien species invasions. 4

- For an overview of the alien species problem in Hawai'i, visit
  www.hawaii.gov/dlnr/Aliens3.html
  www.botany.hawaii.edu/botany/news/silent.htm

- Don't bring to Hawai'i any food, animal, or plant matter that could contain live organisms.

- Never dispose of a pet by releasing it into nature.

- If you have information regarding illegal animals in Hawai'i, call the state's Toll-Free Pest Hotline at 643-PEST(7378).

- Inform anyone with an illegal pet (including snakes) about the state's penalties and amnesty policy.
  www.hawaiiag.org/hdoa/pi_pq_amnesty.htm

- When you arrive in Hawai'i from somewhere else, check your suitcases, boxes, and even the soles of your shoes for stowaway insects, plant seeds, etc. If you find some, destroy them.

- Buy native plants grown in Hawai'i.

- Learn to identify Hawai'i's native creatures.

- Hike through the forest or snorkel with a guide familiar with Hawai'i's native creatures and the way they were revered and used traditionally by the Hawaiians.

PRESERVE THE PRISTINE PLACES

5 Reduce the number of feral animals.

Roughly 22,000 cats are euthanized each year in Hawai'i. Unwanted dogs and rabbits are also euthanized. Many cats, dogs, and other pets are abandoned and become feral, often damaging Hawaiian ecosystems by competing with or eating native species. (Feral cats are notorious for eating birds.) Feral animals often suffer from hunger and disease. The best ways to decrease the number of feral animals are to

- **Neuter your dog or cat as soon as possible.**
www.hawaiianhumane.org/programs/neuternow/index.html

- **Become a cat colony keeper if you have feral animals in your neighborhood.** A colony keeper has all of the animals in the colony neutered for a very low price and then feeds them so that they live out their lives under humane conditions without reproducing.
www.hawaiianhumane.org/programs/feralcat/index.html

- **Refer anyone with an unwanted pet to the Humane Society.**

Don't support poaching. 6

Poaching occurs when a creature is harvested illegally. Poached creatures or their parts are often sold on the open market. If there is no official certification system in place, the buyer has no way of knowing the creature has been harvested illegally.

- Ask to see certificates, check their authenticity, and buy only certified items.

- Don't buy parts of protected or endangered species: ivory, rare shells, turtle shell. (It is illegal to sell endangered species products.) For a list of Hawai'i's endangered species, visit

hbs.bishopmuseum.org/endangered/endangeredhi.html

- Don't buy anything that was part of an ocean-dwelling organism unless you're going to eat it or know it was grown in aquaculture or taken as a part of a sustainable fishery. If you don't know, assume the worst!

- It is okay to buy these certified marine organisms: cultured pearls, mother-of-pearl items from cultured shellfish, faux pearls, faux shells, faux coral, abalone jewelry made from cultured abalone, shell jewelry in which the shells were collected when the animal was already dead (Ni'ihau shell jewelry, some "*puka* shell" jewelry, anything made from a cultured marine organism).

www.scscertified.com/fisheries/

- Absolutely never buy live saltwater fish that are not certified, dried sea stars, dried seahorses, real shells,

PRESERVE THE PRISTINE PLACES

coral skeletons, sea urchin skeletons, sand dollar skeletons, sea urchin spine decorations or jewelry, dried fish (unless you eat them), abalone items from wild-caught abalone, anything made from undersized *'opihi* (limpets).

- Think twice about buying precious coral jewelry, mother-of-pearl jewelry, wild-caught pearls.

- If you ever see anyone in Hawai'i catch, harass, or kill an endangered or otherwise protected species, call the U.S. Fish and Wildlife Service office nearest you.
www.fws.gov/offices/directory/

Forest and Kim Starr

Know where things you buy come from. 7

If a product is made in the United States or other developed country, the resource extraction and labor conditions may be environmentally sound and humane. While a good argument can be made for supporting developing countries by purchasing their products, we may be doing the people of those countries a disservice by supporting poor working conditions and extraction methods that are ruining their environment.

- **For a list of certified environmentally responsible products, visit**

  www.scscertified.com/manufacturing

  www.mbdc.com/certified_levels.htm

- **Look for lumber that comes with certifications that guarantee that the wood was harvested sustainably.**

  www.fscus.org/

PRESERVE THE PRISTINE PLACES

PRESERVE THE PRISTINE PLACES

8 Calculate your ecological footprint and shrink it.

Your "footprint" is an estimate of how much of the surface of the planet is needed to support you and your lifestyle for a year. It's a quasi-quantitative measure of how heavily we use the Earth's resources.

- **To calculate your footprint, visit**
  www.myfootprint.org
  www.bestfootforward.com/footprintlife.htm

- **Make and carry out a plan to decrease your footprint's size.**

Reproduce responsibly. 9

The "replacement level" fertility rate in developed countries is 2.1 children. (Every tenth couple can have three; hence the 2.1). For a hard look at the population issue, visit

www.npg.org

www.populationconnection.org

PRESERVE THE PRISTINE PLACES

10 Support conservation-oriented policy changes.

PRESERVE THE PRISTINE PLACES

- Testify before the state legislature. Look up the details about any bills before the legislature and find out when testimony can be given by visiting

  www.capitol.hawaii.gov/site1/docs/docs.asp

- The Sierra Club produces a Legislature Environmental Scorecard each year that rates the performance of our state senators and representatives based on the environmental bills they voted for or against. For the names of legislators to lobby for your favorite environmental cause, visit

  www.hi.sierraclub.org/scorecards/

- Attend a Neighborhood Board or other public meeting and speak up about preserving the pristine places you care about. Find out anything you want to know about your Neighborhood Board at the Neighborhood Commission's Web site:

  www.honolulu.gov/nco
  Phone: (808) 527-5749

- If, as a visitor to Hawai'i's parks, reefs, and forests, you have a chance to offer written comments that will strengthen the protection of a place you love, take the time to fill out the form and talk to the people who work for the protected place. Show your support.

Don't contribute to polluted runoff. 11

- Don't cover your property with concrete or other impervious surfaces.

- If you must cover it with something, use water-permeable concretes and asphalts. They allow water to drain through them and follow its natural path to the ocean without becoming polluted by dirty impermeable surfaces.

- Let there be dirt!—with plantings.

- Avoid fertilizing your yard or potted plants. If the extra nutrients get into rivers and streams, they cause excessive algae growth, which smothers and kills coral and is toxic to humans if the algae are species that create toxins.

- If you have a septic tank, have it checked for leaks and maintain it well. For information on maintaining your septic tank, visit hawaii.uscity.net/Septic_Tanks

- Report a clogged storm drain to your City and County Department of Environmental Services. For O'ahu numbers, visit

  www.honolulu.gov/env/wwphones.htm

- Report any situation that may pollute storm drains, waterways, and beaches. On O'ahu call the Environmental Concern Hotline at 692-5656.

KEEP CORAL REEFS HEALTHY

- Don't flush toxic cleaners and other products down the toilet or shower. They may escape into the soil through leaky sewer pipes. Even if they make it to the waste treatment plant, the plant may not be equipped to clean up those toxic chemicals.

- Check your vehicles for leaks and fix them. If you cannot fix them, absorb the leaks with cardboard or other absorbent material placed under your car.

- Wash your car at a car wash that recycles the water, or try a dry wash (but check to see how toxic it is). Use a biodegradable, low-phosphorous detergent if you must wash your car yourself. Wash it as far from any body of water as possible. Consider never washing your car at all! (This may be impossible for many and an easy "sacrifice" for others.)

- When you wash your boat, use just fresh water. If you must use detergents, make sure they are nontoxic and low in phosphorous.

Prevent marine debris. 12

Hawai'i is famous for being the "comb of the Pacific." The archipelago is situated so that it collects marine garbage from all over the Pacific Rim. All the trash you see washed up on our beaches is a form of pollution called "marine debris." Most of it is plastic. It is deadly to the sea birds, whales, and sea turtles that ingest it. It can also strangle sea life and is a hazard to marine vessels. Major reductions in marine debris will require strong international laws. In the meantime, there is much you can do.

- Participate in International Coastal Cleanup Day through Hawai'i's annual Get the Drift and Bag It campaign. Every September, beaches all over the world are cleaned up, and the collected debris is documented, enabling governments and scientists to monitor changes.

www.coastalcleanup.org/pub1/

KEEP CORAL REEFS HEALTHY

- Don't take to the beach plastic bags, straws, paper cups, or anything else that can blow or float away and end up in the gut of a turtle or sea bird.

- If you see a bag, a balloon, six-pack rings, or any other nontoxic litter at the beach, pick it up and throw it away. When swimming, snorkeling, diving, spearfishing, or canoeing, collect trash, tangled fishing line (watch out for hooks), or abandoned fishing nets.

- Avoid setting off fireworks at the beach, since the paper and chemicals on the paper end up in the ocean. If you do set off fireworks at the beach, pick up the trash that remains.

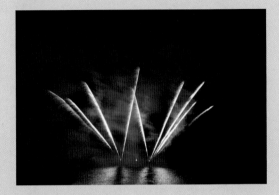

Protect our shorelines. 13

- If you live on the shore (or next to a river) educate yourself to avoid dangers to your home or the reef. Visit the State of Hawai'i Department of Land and Natural Resources Office of Conservation and Coastal Lands Web site:

  www.hawaii.gov/dlnr/occl/index.php

- Know how rapidly the shoreline is changing.

- If you're building, use an ultrasafe setback to avoid future problems.

- Don't build a seawall; it accelerates beach loss, changes the health of reefs, and potentially reduces sea turtle nesting sites.

14 Fish responsibly.

USE THE OCEAN WISELY

- Don't use illegal gear, fish in prohibited areas, or catch and keep small fish and sea creatures. Doing so condemns future generations to having fewer fish than we do.

- Speak up if friends and family catch and keep fish that are too small (those that haven't had a chance to reproduce).

- Teach your children how to fish responsibly. Pass on the traditional knowledge the Hawaiians had about how to manage ocean resources. Their *kapu* system had a purpose. If you don't know the old ways, learn them or use modern equivalents.

- Be familiar with Hawai'i's fishing regulations, which change frequently. They are listed on the Department of Land and Natural Resources Division of Aquatic Resources (DAR) Web site:
www.hawaii.gov/dlnr/dar/fish_regs/index.htm

- If you don't fish, you should still know the regulations. If you see a violation, report it, anonymously if you wish, on O'ahu: 587-0077; Hawai'i (Hilo): 974-6208; Hawai'i (Waimea): 887-6196; Hawai'i (Kailua-Kona): 327-4961; Hawai'i (Capt. Cook): 323-3141; Kaua'i: 274-3521; Maui: 873-3990; Moloka'i: 553-5190; Lāna'i: 565-7916. There is an informer's fee opportunity: half of the fine collected as a result of conviction for a violation may be paid to the informant.

Eat seafood responsibly. 15

Even if you do not fish, you can prevent overfishing and support fisheries that are fished sustainably.

- Learn what kinds of seafood in supermarkets or restaurants are not overfished. Two agencies providing online and snail mail lists of safe seafood are Seafood Choices and Blue Ocean. To find out which fish are currently okay to buy or order, visit their Web sites:

www.seafoodchoices.org/smartchoices.php

blueocean.org/seafood

Neither agency focuses on Hawai'i's fish, however. We need a listing of the status of all the fish caught and sold off Hawaiian waters.

- Ask servers or store butchers if they know whether the fish they sell are aquacultured. A number of fish species are aquacultured in Hawai'i. These include Pacific threadfin *(moi)*, tilapia, catfish, carp, flounder, sturgeon, amberjack, snappers, and grouper. Not all aquaculture is better for the environment. If the items are grown in Hawai'i they are subject to more stringent regulations than if they are grown in a less developed foreign country.

USE THE OCEAN WISELY

16 Live simply.

- Buy only what you really need—and a few things you REALLY want.

- When you think you need some "retail therapy," go for a walk with a friend instead, or go surfing or jogging.

- Buy fewer, smaller, smarter, better quality, less impactful things.

- Adopt the philosophy that less is more, smaller is better.

- If you're worried about America's Gross National Product or the Gross World Product, support the economy in a way that doesn't produce potential solid waste by buying

  - massages, facials, restaurant gift certificates, or concert tickets.
  - items made in the region you live in (to eliminate shipping, i.e., fossil fuel use, and to support the local culture and economy).
  - items that will last a long time and that may even have a reusable afterlife.
  - items that are recyclable.
  - items made from recycled materials.
  - items that are biodegradable.
  - items manufactured or extracted from nature in a way not harmful to the environment.

Avoid overpackaged and disposable items. 17

Recycling, although it is a better option than landfilling or incinerating trash, still requires energy and new resources to remanufacture goods with the recycled materials. Without becoming a pack rat, reduce the amount of trash you produce:

- When you food shop, buy the least packaged brands. Minimize the number of things you have to open before getting to the product.

- Avoid subdivided goods. Instead, buy in bulk and subdivide food in reusable containers.

- Send your children to school with reusable lunchboxes or bags and put food in reusable containers that they learn to bring home.

- Use cloth napkins instead of paper napkins.

- Use reusable dishtowels instead of paper towels.

- At picnics and luaus
  - use reusable plates, utensils, and cups.
  - pack foods that require few plates or utensils.
  - aim to picnic without generating any waste at all, except for organic leftovers and recyclables.

- Compost your organic (food) leftovers and use the compost as fertilizer in your yard or garden. For tips on composting and recycling home organic wastes, visit
www.recyclehawaii.org/
www.hawaii.gov/health/environmental/waste/

REDUCE SOLID WASTE

18 Keep your drinking water clean and simple.

REDUCE SOLID WASTE

- Determine if you can drink water right out of your tap by visiting
www.epa.gov/safewater/dwinfo/hi.htm

- Test your home's water quality with any number of easy-to-use and inexpensive water-quality test kits.
www.h20kits.com/

Call the Safe Drinking Water Hotline, 1-800-426-4791, if you have questions.

- Use a filtering system to filter tap water into a reusable container. This will save you money and save the energy needed to recycle disposable water bottles.

- If you have a choice between buying a large or a small water bottle, buy the large one, because it contains less plastic per volume of water.

Go paperless. 19

Going paperless may seem impossible, but see how close you can come to this goal.

- Use a dry-erase board or chalkboard for messages.

- Use a device that lets you make paperless notes. Many of us do this on our cell phones with text messages.

- Use printed-on-one-side waste paper for notepaper and then recycle it.

- Send electronic memos at work.

- Don't print out e-mails and documents; save them in a safe electronic place.

- Read online newspapers and magazines. If you must buy printed magazines, recycle them (after everyone who wants to read them has had a chance).

- When you finish reading books, keep them for future readers, donate them to your local library, or sell them to a used bookstore near your home.

REDUCE SOLID WASTE

20 Reuse.

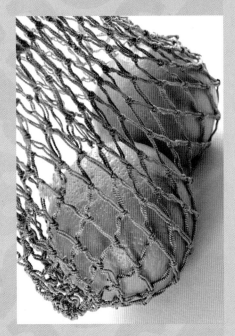

Many items have more than one use and more than one "life." Reuse saves you money and saves the 'āina by filling up the landfill more slowly.

- Reuse plastic and paper bags.

- Take your own bags to stores. If a store won't let you use your own bags, complain. Then shop somewhere else, or at least reuse or recycle the store's bags.

- Consider items you own as things with potential, not trash, and devise new uses for them. For creative solutions for reusing things, visit

www.make-stuff.com/recycling/

www.redo.org

Repair. 21

REDUCE SOLID WASTE

- Don't even think of throwing away anything until you look in the phone book for a repair shop. Shoes, televisions, clothing, microwave ovens, and ukuleles—all of these and many more can be repaired for less than it costs to buy new ones.

22 Recycle bottles.

REDUCE SOLID WASTE

- To learn how and where to recycle bottles in Hawai'i and redeem your 5¢ per container, visit

  www.hawaii.gov/health/environmental/waste/sw/hi5/index.html

23. Recycle aluminum, plastics, white paper, newspaper, paper bags, cardboard, and glass.

- To learn where to recycle the above items, visit

Oʻahu:
envhonolulu.org/solid_waste/

For curbside recycling on Oʻahu:
ocr2000.com

Kauaʻi:
www.kauai.gov/recycling

Big Island:
www.recyclehawaii.org/where.htm

Maui, Molokaʻi and Lānaʻi:
www.co.maui.hi.us/departments/Public/Recycle/index.htm

For curbside recycling and other information on Maui:
www.mauirecycles.com/

For the Maui Recycling Guide:
mauirecyclinggroup.org/GuideHome.htm

To buy recycled items on Maui:
mauirecyclinggroup.org/remade_on_maui.htm

24 Recycle the odd stuff.

REDUCE SOLID WASTE

- Search the Internet to find where to recycle odd items. For example, if you want to recycle green waste on Oʻahu, search for "Recycle green waste Oahu."

- For comprehensive information on recycling, visit

  envhonolulu.org/solid_waste/Greencycling.htm

Some items you can recycle in Hawaiʻi:

- Car batteries.
  Retailers are required to recycle your old battery when you buy a new one. You can also take your old battery to City Convenience Drop Off Centers. For the Centers in Honolulu, visit

  envhonolulu.org/solid_waste/

- Tires.
  These also may be taken to City Convenience Drop Off Centers.

- Large appliances.
  These also may be taken to City Convenience Drop Off Centers.

- Green waste.

  envhonolulu.org/solid_waste/Greencycling.htm
  www.kauai.gov/Default.aspx?tabid=113
  www.recyclehawaii.org/where.htm
  www.recyclingtoday.com

- Computers.
  Donate functional computers to the Hawaiʻi Open Source Education Foundation (HOSEF). Call 689-6518 or visit

  www.hosef.org

  Recycle dysfunctional computers. Pacific Commercial, Pacific Environmental Corporation, Philip Services, Enviro Services, and Haztech handle e-waste for a fee.

- Batteries.
  www.mauirecycles.com/recycle_batteries_inkcartridges.htm

- Ink cartridges.
  Office Max, Office Depot, and other office supply stores recycle cartridges. For another option, visit

www.mauirecycles.com/

- Metals other than aluminum.
  On Oʻahu put tin/steel cans in with your household trash. Your residential trash is taken to H-POWER, the city's waste-to-energy plant. Mechanical separators pull ferrous and nonferrous metals from the trash. The metal is cleaned and sold to a metal recycler. For other details for Oʻahu, visit

www.hawaiimetalrecycling.com/Acceptance.htm

- Motor oil.
  Used motor oil, cutting oil, or fuel oil can be disposed of with your regular household trash. Use an oil change box or pour the oil into a plastic bag with an absorbent material and seal the bag. Commercial generators must handle and dispose of used oil in accordance with EPA and State Department of Health guidelines. For more details, visit

www.recyclehawaii.org/oil.htm

- Cell phones.
  For directions to a nonprofit company that will donate a portion of its recycled cell phone and battery sales to a partnered nonprofit organization, visit

www.call2recycle.org/

- Telephone books.
envhonolulu.org/solid_waste/
www.co.maui.hi.us/departments/Public/Recycle/

- Christmas trees (after Jan. 1).
www.hawaii.gov/health/about/pr/pr/2003/03-99trees.html

REDUCE SOLID WASTE

25 Give away or sell unwanted items.

- Give them to a family member or friend.

- Have a garage sale.

- Donate them to a thrift shop.

- Donate them to Goodwill, United Cerebral Palsy of Hawai'i, or other charitable organizations.
  www.higoodwill.org
  www.ucpahi.org

- Give them to persons in need. Be sure to ask them first.

- List them on a Web site made just for people who want to share their used things with others:

  On Maui:
  www.alohashares.org

- Place a free ad for a giveaway on the Internet.
  honolulu.craigslist.org
  www.freecycle.org

Don't generate hazardous waste. 26

- Avoid buying products that generate household hazardous waste. For a list of products and other useful information, visit

  www.epa.gov/region09/waste/solid/house.html
  envhonolulu.org/solid_waste
  www.ctahr.hawaii.edu/oc/freepubs/

- Use less dangerous alternatives to hazardous products. For lists of nonhazardous and often money-saving products, visit

  www.epa.gov/region09/waste/solid/house.html
  www.nontoxic.com/nontoxic/askdoctor.html

- When there is no substitute product, use the hazardous product properly and use it all so there is nothing left over to dispose of. If you cannot use it all, offer it to a neighbor or a charitable organization. Call HIMEX, the Hawai'i Materials Exchange, at 586-8143 to list usable products in a statewide database.

- If you cannot get rid of unwanted leftover toxic household substances, dispose of them properly. For a general disposal guide, visit

  www.hawaii.gov/health/environmental/waste/sw/index.html
  envhonolulu.org/solid_waste

Three islands have designated dates and locations for dropping off or disposing of unwanted toxic substances safely:

O'ahu:
Call Household Hazardous Waste Disposal: 692-5411

Big Island:
www.recyclehawaii.org/calendar.htm

Kaua'i:
www.kauai.gov/Default.aspx?tabid=114

KEEP YOUR HOME HAZARDOUS-WASTE FREE

27 Use less water inside.

- Don't keep the water running when you're washing dishes, brushing your teeth, shaving, or washing produce.

- When you wash clothes or dishes, wash full loads and use cold water for clothes whenever possible to save energy.

- Check faucets and pipes for leaks.

- Install water-saving showerheads and faucets. Two popular types are lower flow (also called restricted flow) and aerated flow. Lower-flow fixtures reduce the amount of water by letting a smaller steady stream flow through the faucet or showerhead. Aerated-flow fixtures add air to the flow so it looks and feels as if more water is flowing than really is.

- Decrease the volume of water used when you flush. Buy a displacement device to put in the toilet, install a toilet dam, have a plumber decrease the volume of water by adjusting your toilet, or buy a low-flush toilet. If you can get your entire condominium to change to low-flow toilets, the whole building uses less water and your maintenance fees may come down.

- Take showers instead of baths.

- Take shorter showers.

- How about a Hawaiian rinse in our ocean waters (which you've helped to keep clean), followed by letting the salt dry on your skin. It's a natural spa experience!

- For 101+ more ways to save water, visit
www.h2ouse.net
www.eartheasy.com/live_water_saving.htm
www.boardofwatersupply.com

SAVE WATER

28 Use less water outside.

SAVE WATER

- To reduce watering, plant native species adapted for the climate in your area.

- If you must water your yard plants, do so early in the morning or late in the day. Because less water will be lost to evaporation, you can water for a shorter period.

- Minimize lawn areas. Native Hawaiian coastal zone ground cover preserves Hawai'i's uniqueness and needs no watering.

- Xeriscape. This means to landscape with plants that require little or no watering. Many Hawaiian plants are suitable for a xeriscaped garden, which can include gravel or crushed lava around the plants. For general information on xeriscaping, visit www.xeriscape.org
www.greenbuilder.com/sourcebook/xeriscape.html

On O'ahu visit the Halawa Xeriscape Garden, 99-1268 Iwaena Street, 'Aiea, HI 96701, (808) 748-5041.

- Put the right plant in the right place. Pay attention to how sunny or shady each area of your yard is and how moist the soil is naturally. Plants on the south and west sides of a building are often exposed to more hours of sunlight and higher temperatures. Plant heat- and dry-tolerant plants in these areas.

- Mulch. Covering the soil with mulch helps it stay moist and stay put, and may add nutrients while keeping weeds from germinating. Mulches include bark chips, crushed lava rock, lawn shavings, and gravel. The County of Hawai'i has FREE MULCH available seven days a week from 6:30 A.M. to 6:30 P.M. at the public pickup area of the Kailua-Kona (Kealakehe) Transfer Station. For places on O'ahu that have free mulch for pickup, visit envhonolulu.org/solid_waste/Greencycling.htm

  For other locations, search the Internet for "free mulch" followed by the name of your town or island.

- For more ways to conserve water outdoors, visit
  www.boardofwatersupply.com/
  www.cabq.gov/waterconservation/outdoor.html

Heidi Bornhorst

29 Burn calories, not fossil fuels.

Every time you use your own energy to get somewhere, you save yourself the cost and the planet the pollution caused by using another form of energy. Here are some ways you can burn your own fuel instead of fossil fuels:

- Walk.

- Bike.

- Take the stairs, not the elevator or escalator.

- Instead of driving around in circles looking for the perfect parking space, take the first or farthest one you see. You'll get some exercise, and your car is less likely to get dinged if you park farther away, where fewer others do.

- Pass these habits on to your children.

Use lighting efficiently. 30

In Hawai'i 95% of the energy used in homes and businesses comes from burning fossil fuels (oil or coal). There are many ways you can decrease the amount of energy your place consumes. If you pay your own electric bill, all of the following will save you money. If you don't pay your electric bill directly because you live in a condominium or apartment, or are just visiting Hawai'i, you will save your condominium association, landlord, or hotel money, and this may trickle down to you. In any case, you will be decreasing Hawai'i's dependence on oil and coal, and you will be decreasing the amount of greenhouse gases Hawai'i emits.

- Turn out the lights when you're not using them. Teach your children to do this too.

- Buy high-efficiency, low-energy, long-lasting bulbs. Compact fluorescent light bulbs are widely available now. Although they cost more up front, they use less energy for equally bright light and last much longer than regular incandescent bulbs.

- Look for the Energy Star seal whenever you buy light bulbs or fixtures. Energy Star, a joint program of the U.S. Environmental Protection Agency and the U.S. Department of Energy, identifies products that meet energy efficiency guidelines. For a list of of energy-saving appliances and tips, visit

   www.energystar.gov

- Dust your light bulbs. They will provide more light, allowing you to use less additional lighting.

- Install dimmer switches so you can tailor the light level to your activities.

- Use natural daylight when you can.

31 Cool your home efficiently.

- Take advantage of Hawai'i's trade winds. If you're building a house, plan a layout and choose the types of windows that let in cooling breezes.

- Plant trees around your house. Trees provide shade and absorb carbon dioxide.

- If you must use air conditioning:

  - Buy energy-efficient window units. They allow you to cool only the rooms in use. Look for the Energy Efficiency Rating (EER) or an Energy Star seal and choose a size that is right for the room. For discount coupons for energy-efficient air conditioners, visit www.heco.com

  - Make sure the seal around a window unit is tight.

  - Keep the filter clean.

  - Set the thermostat for window units or central air at 78 degrees or higher.

Buy energy-efficient appliances. 32

- Buy only appliances with an Energy Star seal or some other verification that they use less energy than others like them and are efficient with the energy they do use. BOTH are important: low energy use and high energy efficiency.

- When you're not using appliances for long periods, unplug them. Many appliances use electricity to run a digital clock or some other internal part.

CONSERVE ENERGY

33 Reduce appliance use.

- Keep only the appliances you absolutely need. Do you really need an electric nail dryer? An electric air freshener? You'll have more space if you get rid of them; you can dry your nails on the lanai and open a window to freshen the air. Recycle discarded appliances if possible.

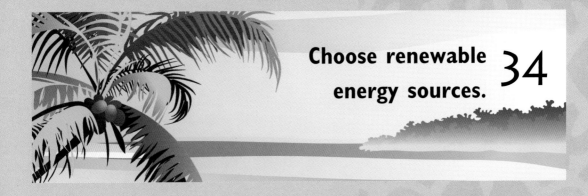

Choose renewable energy sources. 34

- Install a solar water heater. In Hawai'i you get state tax credits and up to a 40% rebate on the cost of solar water heaters. Using solar can cut your water heating costs by 90%, saving you up to $10 per person per month.

- Install a photovoltaic solar panel or panels. This will provide you with free electricity (after you have paid for the cost of the solar panels).

- Install a home wind turbine!

www.awea.org/smallwind/hawaii_sw.html

- Buy a home that has an Energy Star seal of approval or is Leadership in Energy and Environmental Design (LEED) certified.

www.usgbc.org/DisplayPage.aspx?CMSPageID=147

These homes are designed to use energy sparingly and efficiently and are constructed with environmentally friendly materials.

- For tips on building an energy-efficient, water-conserving house made of products that are Earth friendly, visit

mts.sustainableproducts.com/modules/
www.dreamgreenhomes.com
www.thegreenguide.com/green-homes
www.sustainablehomes.co.uk

35 Perform an energy audit.

An energy audit is like a physical exam for your home. Taking into account the age of the building, its size, its location, and your appliance use, an audit can show you how to save money by using less electricity. You can do your own home energy audits; just follow the directions at either of these sites:

hes.lbl.gov

www.eere.energy.gov/consumer/

Lobby for more and better energy bills. 36

- Attend any public meetings in your neighborhood that relate to energy and lobby for anything that reduces Hawai'i's energy needs or increases the amount of energy that comes from renewable energy forms.

- Testify for energy-saving bills. For details about any bills before the legislature and times testimony can be given, visit www.capitol.hawaii.gov/site1/docs/docs.asp

CONSERVE ENERGY

37 Drive less.

REDUCE GLOBAL WARMING EMISSIONS

- Live near where you work or work near where you live.

- Carpool.
  www.vanpoolhawaii.com

- Perform your errands and do your shopping as close to home as possible. Further shorten the distance you travel by doing errands without backtracking.

- Take the bus or trolley.

- Ride a bike. Because Hawai'i has a shortage of bike lanes, biking in Hawai'i can be dangerous. For tips on safe biking, complete with maps showing bike lanes and unsafe roads, visit
  www.hbl.org/maps/maps.html

- Walk.

- Vary your routine: walk some days, take the bus other days, drive on the days you have errands, ride with a friend or family member when you can.

- Support the development of new mass transit infrastructure and improvements to our existing well-used bus system.

Use less energy when you drive. 38

- Drive the speed limit. Driving faster uses more gas per mile traveled; you will save money as well as energy by driving more slowly.

- Buy a vehicle that gets the best gas mileage in its class.

- Buy a vehicle that gets really good gas mileage, period!

- Buy a hybrid vehicle.

- Look into using alternative fuels: biodiesel, ethanol, or gasoline with as high a percentage of ethanol as possible.

biodiesel.com
www.hawaii.gov/dbedt/ert/cc/ccfsbf.html
www.doe.gov/energysources/bioenergy.htm

- Improve your vehicle's gas mileage:

  – Keep your tires inflated at the proper level.

  – Don't buy tires with excessive tread; you'll literally pay for this "cooler" look with every mile you drive.

  – Take surf, canoe, or kayak racks off the top of your car when you're not using them. They increase wind resistance, which decreases your gas mileage.

REDUCE GLOBAL WARMING EMISSIONS

39 Buy organic food.

The United States has fairly strict regulations governing what can be called "organic" (although the same does NOT hold for the terms "green" or "lite"). The National Organic Program, run by the United States Department of Agriculture (USDA) is in charge of the legal definition of "organic" in the United States and certifies products as organic. According to its definition, "organic foods" are foods produced according to organic standards. These include

- crops grown without the use of conventional pesticides, artificial fertilizers, or sewage sludge;

- animals reared without the routine use of antibiotics and without the use of growth hormones;

- foods processed without ionizing radiation;

- foods processed without the use of a wide range of food additives;

- foods produced on all levels without the use of genetically modified organisms.

Organic foods are often more expensive than nonorganic foods, but they are becoming more affordable as well as more available. The following seals indicate that food is certified organic. Look for these or a seal from the country of origin when you buy organic.

When you buy organic food, you help prevent pollution of our soils, streams, fresh water, oceans, and, inevitably, the air as well.

Prevent our next sewage spill. 40

In 1990 the first edition of this book rightly urged us all to lobby our legislators to upgrade existing sewage treatment. In 2006, a large sewage pipe broke, releasing 48 million gallons of raw sewage into the Ala Wai and onto the coral reef and beaches of Waikīkī. It's time the state heeds this original advice and spends the money needed to ensure that all our sewage makes it to the waste treatment plant. We also need to spend in order to upgrade and expand our waste treatment plants as Hawai'i's population grows and we make more demands on our existing infrastructure.

Here's what you can do:

- Report any sewage spill, leak, or smell to the authorities as soon as possible. On O'ahu: Environmental Concern Hotline, (808) 692-5656. All other islands: Department of Health's Clean Water Branch, (808) 586-4309.

- Testify in support of improvements to your neighborhood's or island's or the state's sewage system.

PREVENT POLLUTION

57

PREVENT POLLUTION

41 Support politicians who improve public infrastructure.

"Infrastructure" includes our publicly owned wastewater treatment plants, energy and water delivery systems, streets, sidewalks and landscaped areas, parks, and beaches. Our infrastructure must be in good functioning order every day to maintain a good quality of life for the taxpayers who support it. Vote for politicians who pledge to take care of the islands' infrastructure.

Learn about the land. 42

- Observe Hawai'i's natural wonders. Pay attention to sunrises and moonrises, sunsets and moonsets, the equinox, the solstices, falling stars, waves, winds, rains.

- When you walk on the beach, a hiking trail, or a street, take a bag with you and pick up nontoxic litter. You will be helping keep Hawai'i clean and, more important, you will be a role model for others.

- Educate yourself about Hawai'i's natural wonders.

  - Take a night class or get a new degree.

  - Learn to distinguish Hawai'i's native plants and animals from those that are not native to Hawai'i.

  - Talk with someone knowledgeable about the native uses of Hawaiian plants and animals.

  - Go fishing with an experienced fisherman or ask your grandfather to show you how to sew a fishing net to responsibly catch a certain fish in a certain place.

  - Go pig hunting with someone who knows the forest.

PROMOTE A CULTURE OF LOVING THE LAND

PROMOTE A CULTURE OF LOVING THE LAND

- Learn how to properly gather *limu* or lei flowers.

- Take a class or go to a talk at the Waikīkī Aquarium, the Honolulu Zoo, or one of the botanical gardens.

- Hula is a dance form that conveys a love and knowledge of nature. Take hula classes or attend a hula performance in which the *kumu* explains what the dancers are expressing.

- Read locally published books about Hawai'i's natural history.

- Eat a Hawaiian meal made completely of traditional food. Learn how the taro was grown and turned into poi, where and how the fish was caught, how the *lū'au* leaf is prepared for the *lau lau,* and where the pig was hunted.

- Take up a hobby that celebrates nature:
  - Hiking

  www.hawaiitrails.org/home.asp
  - Snorkeling, scuba diving, underwater photography
  - Bird watching

  www.hawaiiaudubon.com/
  - Surfing
  - Pig hunting
  - Native plant cultivation
  - Lei making
  - Gardening
  - Painting
  - Writing poetry

Teach about the land. 43

- Take your friends hiking or snorkeling. The experience may make them more protective of Hawai'i's beauty than anything you can say to them.

PROMOTE A CULTURE OF LOVING THE LAND

44 Eat whole foods.

Processed foods have introduced potential health risks and increased the energy our food requires for manufacture before it reaches us. They also often have much more packaging. Just think of the manufacturing and packaging cost differences between raisins and jelly beans, fresh fruit and store-bought cookies, *poke* and bologna.

Support Earth-friendly and locally owned establishments. 45

- When you patronize a business, look into its practices. Does it recycle? What does it recycle? Does it sell products made from recycled materials? Does it sell goods made from sustainably harvested materials? Does it sell products that are certified environmentally friendly in their manufacture or use? Is the mission or product line good for the Earth?

- Whenever possible, buy Earth-friendly goods and services that are made in Hawai'i.

PROMOTE A CULTURE OF LOVING THE LAND

46 Invest in environmentally responsible companies.

Some mutual funds include stocks for only environmentally friendly companies. Before investing in such funds, be sure to read the fine print. The definition of what constitutes an environmentally friendly fund varies. Make sure the definition is one you're satisfied with.

Don't build or buy in high-risk areas. 47

Cliffs and seashores are dynamic places in a geological sense. They may not change on a daily scale, but they certainly change over time. Don't put your life savings into a place that may end up underwater, down a mountainside, or bombarded by falling boulders. To learn what is happening geologically in coastal areas, visit the Department of Land and Natural Resources Office of Coastal and Conservation Lands Web site:

www.hawaii.gov/dlnr/occl/index.php

The state is currently reviewing the laws that guide developers along the coastlines. The laws may be revised to help protect prospective buyers.

PROMOTE A CULTURE OF LOVING THE LAND

48 Travel responsibly.

Whether you have traveled to Hawai'i or you live here and are planning to go *holo holo* (traveling for pleasure) somewhere else, there are things you can do to help keep Hawai'i or any destination healthier:

- Look for Green Globe 21 vacation destinations.

  www.greenglobe.org

- If you can, stay at places that have been built to LEED design standards. (Search the Internet using the search terms LEED and the names of specific hotels.)

- Take advantage of local artisans and local services to support the economy in the place you're visiting.

- Do the little things that add up to large energy savings: in hotels, don't ask that your towels and sheets be washed every day; eat at places that have reusable tableware and otherwise minimize the generation of trash.

Get involved at work. 49

- Create or get an environmentally friendly job: Don't do any kind of job that goes against your land-loving grain. And don't simply quit so that someone else does your job; change your workplace so that no one has to do the job anymore. If you want a really green job, visit

  www.environmentalcareer.com
  www.ecojobs.com/index.php
  www.ecoemploy.com
  www.theenvironmentsite.org/

- Become the "Green Mentor" at work or home. Here are a few ways you can be a catalyst for change at work:

  — If you work with toxic substances, find out if less toxic substitutes are available. If they are, ask your boss to change to the safer versions. If the safer versions are more expensive, remind your boss that protecting worker safety may prevent workers' compensation claims in the future.

  — Start a workplace recycling program. Post reminders to turn the lights out.

50 Get involved in the community.

- Run for office.

- Volunteer for any one of Hawai'i's 100+ environmentally oriented charitable organizations. Almost all of them are looking for help! Visit www.malamahawaii.org

PRISTINE PLACES

Pristine places are wilderness places where people's adverse effects on nature are small. Nature is healthy when ecosystems are generally intact and as productive and biodiverse as they would be in the absence of people.

Pristine wilderness has been diminished all over the planet with the growth of human populations and enterprises. Only an estimated 27% of Earth's habitable land remains undisturbed by humans. Humans have altered pristine places by directly overexploiting their creatures and natural resources, by intentionally and unintentionally introducing invasive "alien" species and pollution, by creating a built environment on top of them, by converting them to agricultural use (both plant farming and livestock keeping), and by intentionally altering them with fire and other means for a variety of purposes.

Hawai'i's habitats, particularly coastlines and lowland forests, have been greatly altered by human activities. Half of Hawai'i's people live within five miles of the water's edge. Much of Hawai'i's unbuilt low and slightly sloped lands have been altered by farming and livestock.

Hawai'i has more endangered species than any other state, even though its area equals less than 1% of the total U.S. land mass. Hawai'i's 297 endangered plant taxa compose roughly 30% of the unique native plants in the state. If you hike in Hawai'i, roughly one of every three species you'll see is threatened with extinction.

About 107 unique species of birds lived in Hawai'i prior to the arrival of humans. Today barely 33% remain, and 66% of those are threatened with extinction. Hawai'i's only two native mammals, the monk seal and the

One thing we do know is that aggressive nonnative plants and animals are doing too well in Hawai'i. Roughly 75% of Hawai'i's original species have been replaced by nonnative species, including 5,047 species of alien terrestrial and aquatic organisms, many of which were intentionally brought to the islands, and 343 alien marine species, most of which were unintentionally introduced. At last count, 80% of the fish and crustacean species in O'ahu's streams, 60% of those in Kaua'i's streams, and 67% of those in the Big Island's streams were alien species.

Hawaiian hoary bat, are on the endangered species list. (A third, another bat species, is now extinct.) Although relatively fewer species of reef fish are endangered, the number of species found around the coastlines of the main islands has declined to, at most, 25% of what it was 100 years ago.

We know very little about the status in Hawai'i of animal species without a backbone, fungi, "lower" plants, bacteria, and other single-celled organisms, since they have been less well studied.

CORAL REEFS

Coral reefs are often referred to as the "rain forests of the sea," because they are highly biodiverse. Unlike rain forests, however, reefs often appear to have more animals than plants. If the reef is healthy, the coral animals themselves appear most abundant, followed by many clearly visible types of fish and marine invertebrates. But the plants on the reef are partly concealed from view. We can see large algae species clinging to rocks, but most of the remaining algae are tiny and live inside the coral animals themselves. The algae benefit by getting a safe home and the nutrients they require from the coral, and the coral benefits from the sugar the algae make through photosynthesis.

This relationship between coral and their symbiotic algae partly explains why coral reefs are fragile ecosystems. The coral animal has a thin outer skin, or epidermal layer, that allows sunlight to penetrate the coral's tissue and reach the algae. This design maximizes the sunlight the coral and algae receive, but it also makes it easier for harmful substances to get into the coral. Those substances include excess dirt from erosion, fertilizers, sewage, and chemical pollution from land-based activities, all of which can weaken and kill coral.

Corals can also be harmed by natural phenomena, like hurricanes, exposure to air at low tides, and freshwater inundation. However, since these phenomena have been occurring for a very long time, the corals have developed mechanisms to survive them. Permanent damage results, however, from elevated sea surface temperatures associated with global warming or El Niño events, which can cause coral to lose their algae symbionts and become "bleached."

Bleached corals weaken and very often die.

The development of coastal areas almost always leads to an increase in impervious surfaces—concrete, asphalt—which eventually become coated with substances that run onto the reef during rains. This "nonpoint" source pollution poses a challenge because the toxic substances enter streams and storm drains at too many sites to clean up using current technologies.

In addition to physical stresses, biological stresses can also damage coral reefs. Some 500 million people depend on coral reefs for food as well as coastal protection and tourism dollars. Thirty million of the world's poorest people depend entirely on coral reefs for food. Unfortunately, in many places people have overcollected algae-eating reef fish that are important to the coral because they crop algae that may otherwise smother the coral. Sea urchins, also important on the reef as algae-eaters, have also been overcollected or have suffered disease epidemics, perhaps as a result of stress caused by human alterations to their environment. Finally, coral reefs, like terrestrial ecosystems, are now being inundated

with alien species. Some of these outcompete or harm the native coral species.

Globally, 20% of the world's reefs have been destroyed and show no immediate prospects for recovery. Another 24% are at imminent risk of collapse as a result of human pressures, and another 26% are under long-term threat of collapse. The good news is that Hawai'i's reefs are relatively healthy. Only 8% of our reefs can be described as either destroyed (1%), critically threatened (2%), or threatened (5%). Hawai'i's coral reefs have recently been valued at $10 billion, with direct economic benefits of $360 million per year. Nearly 52% of visitors to Hawai'i dive or snorkel. These activities outrank all other activities, with the exception of just hanging out at a beach.

Hawai'i has multiple varieties of "marine protected areas," but only about 1% of the coastline is fully protected from collection. Research has shown that for effective protection, roughly 25% of the coastline must be heavily protected by law or by community enforcement. As of July 2004, Australia protects 33.3% of the Great Barrier Reef. Hawai'i may well come up with a similar publicly supported plan. So far, however, there has not been enough public support for increasing the extent of protected areas around the main islands except in a few Hawaiian communities.

Fortunately, the public and political will to protect the coral reefs living on and around the Northwestern Hawaiian Islands is greater. The State of Hawai'i and the federal government worked together in 2006 to establish the Northwestern Hawaiian Islands Coral Reef Ecosystem National Marine Monument. Its primary purpose will be to protect this rich and lovely public trust into the future. What we will do on the main islands remains to be seen.

OCEAN RESOURCES

People in Hawai'i love ocean resources. We love to fish for them, eat them, collect them, talk about them, gaze at them, make art from them, tell stories about them, and name watercraft after them. In fact, we love our ocean resources so much that we're at risk of loving them to death.

Those who fish for sport, fish commercially or personally for food, collect fish for aquariums, or collect shells may collect these resources into oblivion. Prime examples of overfished species (species that are collected at a rate greater than the rate at which they replenish themselves) are 'opihi, lobster, Kona crabs, aquarium fish, reef fish, mollusks with beautiful shells (e.g., Triton's trumpet, tiger cowries), limu, bottomfish, tuna, marlin, and swordfish. As collectors become aware of the scarcity of these resources, they begin to take action to protect them. (The National Audubon Society, the United States' oldest environmental organization, famous for protecting birds, was started by duck hunters who were worried about seeing their resource disappear.)

The State of Hawai'i has jurisdiction over its waters from the shoreline to three miles out (except for special protected areas managed by other agencies). The federal National Atmospheric and Oceanic Administration's Western Pacific Regional Fisheries Management Council (Wespac) manages the fishing that occurs from 3 to 200 miles out. (This will change to comanagement in the Northwestern Hawaiian Islands with the establishment of the national monument.)

Because of global overfishing, the oceans may now contain only 10% the number of large predatory fish they once did. Thirty-one percent of the United States' major stocks of fish are overfished. In Hawai'i, the abundance of reef fishes declines as population increases. O'ahu has only one-third the fish biomass of Moloka'i. The Northwestern Hawaiian Islands have 3.6 times that of the main islands. The data clearly show that recreational and commercial fisheries are removing fish faster than the fish can replace themselves. The effects have been so severe in the Northwestern Hawaiian Islands that the commercial lobster fishery has been completely shut down.

We know from archaeological evidence that the early Hawaiians got it right. Their midden piles reflect a long history of eating fish that changed little in size or type over time. Today some Hawaiian communities still manage their marine resources strictly and as a community, and this practice is working. Hanauma Bay, protected since 1967, has many more fish than places where fishing is allowed.

SOLID WASTE

"Solid waste" is officially everything that is discarded as trash and handled as solids, as opposed to the water in our sewer systems, which is officially "liquid waste." Solid waste in developed countries today comes from mining, oil and natural gas production, agriculture, other industries, businesses, government buildings, schools, hospitals, hotels, and homes. Trash from cities and towns is known as "municipal solid waste."

Most trash is not municipal solid waste. Indirect industrial sources, like mining and agriculture, produce 98.5% of all solid waste. Nationally, of this 98.5%, 75% is from mining and oil and gas production, 13% from agriculture, and 9.5% from other industries. While Hawai'i produces its share of agricultural waste and some industrial waste, the amount of agricultural waste produced is decreasing as the sugar and pineapple industries shrink and are replaced by homes or by crops with long-lived plants. Hawai'i produces little or no solid waste from mining and oil and gas production.

We can trace our current trash woes to the industrial revolution, when humans figured out ways to make goods available to booming populations who were staying put with the advent of successful farming. Staying put created the potential to collect things; the industrial revolution made it possible.

In the early days of the industrial revolution, trash was typically disposed of by burying it in a landfill, burning it, or dumping it into a body of water—all methods of disposal that allowed people to think that the trash had gone away. We know that most of it doesn't go away, at least not on a time scale that we care about. Much of it comes back to haunt us in the form of (1) marine debris lying on the beach or permanently stuck in the stomach of a sea turtle or sea bird, (2) air pollution that in some cases returns to the earth when it is combined with water as precipitation, (3) garbage "juice," called "leachate," that can seep through soils into drinking water, or (4) methane gas, a greenhouse effect–producing gas that is created in landfills from billions of methane-producing bacteria dining on our leftovers. These adverse effects, along with the rising costs of land for landfills, landfill disposal "tipping" fees, and transporting trash and the decreasing availability of "suitable" sites for landfills, inspired developed countries to look for new and better things to do with their trash.

Hawaiian tradition favors creating as little waste, or ʻōpala, as possible. Since statehood, Hawaiʻi has mimicked U.S. consumption and waste-generation patterns. The United States has 4.6% of the world's population, but generates about 33% of the world's solid waste. The total generation of municipal solid waste in the United States has increased more than 50% since 1980. In 2005 the average American on the mainland generated 4.5 pounds of municipal waste per day; the average person in Hawaiʻi (including tourists) generated 6.2 pounds per day.

In 1999 (the last year for which a comprehensive study was done) Hawaiʻi generated 1,409,166 tons of trash. Hawaiʻi's population in 1999 represented 0.47% of the total U.S. population, but it generated 0.62% of the total U.S. municipal solid waste. Hawaiʻi's waste contained proportionally more yard waste and wood than mainland waste did for that year.

Hawaiʻi's municipal solid waste in 1999

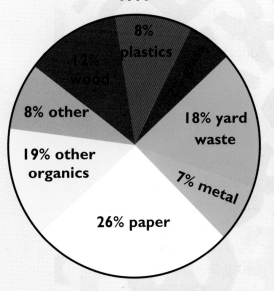

In the United States overall, 56% of our municipal solid waste is disposed of in 1,767 landfills, 14% is incinerated, and 31% is recycled or turned into compost. In 2003, Hawaiʻi landfilled 71% of its waste and diverted 29%. "Diverted" trash is trash that would

have entered the landfill waste stream but was instead collected for recycling, composting, or incineration.

In 1991 Hawai'i developed its first Plan for Integrated Solid Waste Management, with the goal of reducing the state's solid waste stream through source reduction or recycling by 50% by the year 2000. In that year the state diverted roughly 20%, well short of its goal. The 50% goal has since been referred to as "ambitious," which it was (good for us!), and was revised by the EPA to 35% by 2005, another goal we did not achieve.

It is easy to be tough on the state and city and county governments and blame them for not achieving their own goals or the EPA's. Part of the problem is that reducing trash is particularly challenging in Hawai'i:

- We generate more green waste per person than mainlanders do partly because plants grow faster here.
- The tourism industry generates large volumes of trash.
- Most of what we buy in stores is shipped to Hawai'i as packages inside packages inside shipping materials.
- Establishing recycling in Hawai'i poses challenges to commercial recyclers because of the high cost of shipping recyclables to Asia and the U.S. mainland to be reworked and because of the volatility of prices for many recyclable materials.

Nevertheless, we can bemoan the challenges and use them as excuses to keep falling short of our goals or we can galvanize our collective love and respect for Hawai'i and work toward solutions. We obviously cannot continue to create landfill after landfill. O'ahu is having a difficult time finding a future landfill site, while its Waimānalo Gulch landfill is reaching capacity and suffering permit extension after extension. NIMBY (Not in my backyard) is strong in Hawai'i because our neighborhoods are small and vacant land is limited.

Large-scale efforts have or may soon be implemented to reduce our municipal solid waste. The Hawai'i Beverage Container Deposit Program, although still not perfectly implemented, has had documented success. Previously, only about 20% of the 800 million beverage containers discarded annually were recycled. Today we recycle more than 80%. That's 480,000,000 fewer bottles entering the landfills in Hawai'i per year.

A bill recently signed into law requires the City and County of Honolulu to establish an island-wide curbside recycling program by July 2007.

Currently, 7% of O'ahu's electricity is generated from waste incinerated at H-Power. This is enough energy to support more than 45,000 homes. H-Power is not a perfect solution; it generates ash that must be disposed of carefully and also generates carbon dioxide, the major potential cause of global warming. Nevertheless, it decreases the original volume of waste burned by 90%. In addition, H-Power recovers 20,000 tons of metals each year, which are recycled.

A number of more focused measures have been implemented over the years to help reduce the waste stream that enters the landfill bit by bit. Bars and restaurants are required by law to recycle their glass bottles. Office buildings with over 20,000 square feet of office space and government offices are required by law to recycle their office paper. Commercial generators of used paint must dispose of it in accordance with state and EPA guidelines. Curbside-collected yard waste is taken to mulching and composting facilities and businesses. Tires are not accepted at disposal sites. Each county makes recycling available; however, it's not yet as convenient as it could be. We usually have to drive somewhere to drop off our materials.

Hawai'i has not lacked effort or been without good intentions in reducing trash. But we are below average among other states in our below-average nation when it comes to the trash volumes we generate and the percentage of our trash that we recycle.

HAZARDOUS WASTE

"Hazardous waste" is any discarded solid or liquid that is toxic, ignitable, corrosive, or reactive enough to explode or release toxic fumes. In the United States, hazardous chemicals used to be dumped into rivers, injected into deep wells, or placed in surface impoundments and regular landfills. The Clean Air Act of 1970, the Clean Water Act of 1972, and the Resource Conservation and Recovery Act (RCRA) of 1976 changed all that, requiring hazardous waste generators to meet emission and pollution standards and dispose of wastes in very specific, safer ways. Because disposal of hazardous chemicals is often expensive, as are fines for violating regulations, today there are many incentives not to generate hazardous waste, although many in industry argue that the production and use of hazardous chemicals is unavoidable.

The Emergency Planning and Community Right-to-know Act (EPCRA) of 1986 requires all U.S. industries to report to local governments the locations and quantities of toxic chemicals stored at a site and to report any releases of toxic chemicals into the environment.

In spite of these laws and the best efforts of companies and disposal facilities, we are still at risk of exposure to hazardous substances in our daily lives. Ecosystems are at risk, too. Hazardous chemical dumpsites and landfills exist that predate the newer stringent laws. In many cases, these places have been the subject of solid federal cleanup efforts, but in others, cleanups have been inadequate or remain uncompleted. Also, toxic waste spills of all sizes go undetected.

Hawai'i has its share of illegal hazardous waste dumping. Some of this comes from the illegal manufacture of drugs. In other cases, toxic waste generators try to save money by disposing of their waste illegally and insufficiently. In yet other cases, generators hire less expensive but supposedly legitimate hazardous-waste disposal companies that are actually disposing of chemicals illegally. Permitting systems, however, deter this black-market disposal. Whatever the case, when hazardous chemicals are not disposed of properly and are not discovered by authorities, it is only a matter of time until they get into the environment.

Toxic chemicals also enter our environment in runoff from agricultural fields and impervious surfaces, like those on streets, that collect traces of toxic substances that are washed "away" when it rains. When pollutants enter the environment at many small locations like this, they are very difficult to clean up. This kind of chemical pollution is our biggest problem now that we have done much to clean up our major manufacturers and sewage treatment plants. Polluted runoff can end up anywhere. Coastal, stream, lake, and soil ecosystems are notoriously susceptible to damage by toxic runoff.

Households are exempt from the same hazardous-waste disposal laws that govern big business. This does not mean, however, that households don't generate hazardous waste; they just don't generate as much. An incomplete list of substances many of us might have at home that are considered harmful chemicals includes

Disinfectants
Drain-opening chemicals
Toilet bowl cleaners
Spot removers
Latex and oil-based paints
Paint thinners
Pesticides
Weed killers
Ant and rodent killers
Flea powders
Used motor oil
Brake fluid
Transmission fluid
Dry-cell batteries
Swimming pool chemicals

Hawai'i is making progress in reducing hazardous waste: in 1989 the state generated 1,499 tons of hazardous waste; in 2001, the last year for which we have complete data, it generated 781 tons. This reduction shows the result of the higher costs of disposal driving the creation of less toxic alternatives.

WATER CONSUMPTION

The main islands of Hawai'i provide residents with ample clear, clean water, naturally. This is not true of many Pacific islands and many continents. Hawai'i's islands can provide this gift because they are tall enough and positioned in such a way that they create their own rains by drawing water from the moist and nearly steady trade winds. The islands receive some rainfall from passing tropical storms, but the rain that really matters, the rain that the forests depend on, the rain that eventually runs off the mountainsides in streams and rivers or, more typically, percolates down through the surface layers of soil deep into each island's rocky core to be stored, perhaps for years, is the misty rain that falls on mountain slopes day and night.

Whereas we often try to forget about the waste we generate, burying it or shipping it elsewhere, we are more likely to remain aware of the importance of water. The slogans of our county water supply agencies say it well: Water for Life (Honolulu Board of Water Supply); By Water All Things Find Life (Department of Water Supply, County of Maui); Water Has No Substitute . . . Conserve It! (Department of Water, County of Kaua'i); Water Brings Progress! (Department of Water Supply, County of Hawai'i).

Parts of the islands may be dry on the surface, but water exists underground, everywhere. Early Hawaiians were masterful at working with water and the natural watersheds to grow *kalo* and create fishponds. They understood the water cycle, knew the importance of the forests in creating and holding on to the rains, and viewed water as wealth.

Hawai'i's 1.33 million people require water for industry, agriculture, golf courses, residential lawns, showers, flush toilets, car washes, and many more activities. Except for a few short-lived "emergencies" caused by local droughts and temporary infrastructure accidents, Hawai'i has not faced a serious water shortage. However, for all its wealth of fresh water, Hawai'i does not have an endless supply. In some places in the islands, fresh water from streams and reservoirs has occasionally been removed from within the rocks more rapidly than it has been replaced by rain. When this happens, wells produce brackish water.

Before we reach the point of steadily consuming Hawai'i's water more rapidly than it is replaced, we must face some complex questions. What is the best way to make more water available to people at a reasonable cost? How do we best increase water-use efficiency? Are there uses that are wasteful and should they be stopped? At what point are we irreparably harming native Hawaiian ecosystems in our quest for easy water? And finally, is there enough water in Hawai'i to allow us to develop until we resemble Hong Kong? Would we really want that to happen?

Hawai'i has a clean governmental water management system and a State

Water Code that is fairly easy to understand. The Water Code recognizes that "there is a need for a program of comprehensive water resources planning to address the problems of supply and conservation of water." The state's Commission on Water Resource Management sets policies, protects water resources, and defines and regulates water uses.

The complete Water Plan for the state has four components: (1) the Water Resource Protection Plan, (2) the Water Use and Development Plans prepared by each county agency, (3) the State Water Projects Plan, (4) the Water Quality Plan, prepared by the State Department of Health, which addresses what's in our water rather than how much there is.

The goal of the Water Plan is clearly if casually expressed. It aims to document how much water we have throughout the islands, how much we can take sustainably, how much we will need in the future, and how we might meet our needs. The success of the plan depends on accurately modeling just how much water we receive, how rapidly it accumulates, and how much we use.

On the state's Commission on Water Resource Management's Web site, www.hawaii.gov/dlnr/cwrm/data/maps.htm, are maps of each of the main islands showing all the water districts and the amount of water that can theoretically be sustainably drawn from each district daily. For example, the Honolulu district can draw 53 million gallons per day sustainably, while the Lahaina district can draw 40 million gallons per day sustainably.

The state's water situation is currently not at a critical stage (unlike Oʻahu's municipal solid waste situation), but as Hawaiʻi's population continues to grow, water could become scarce if we don't continue efforts to use it more efficiently. Some of the counties recycle municipal waste water and use it in industry and for irrigating crops and golf courses. Although these efforts do cost money, they help conserve water by decreasing the amount of potable water used.

ENERGY

Nearly 90% of Hawaiʻi's energy needs are met by burning oil in some form. Only 30% of the oil we use comes from domestic sources. In 2004, 40% of the nation's energy came from oil. This means that Hawaiʻi is more than twice as dependent as the nation as a whole on mostly foreign oil for its energy.

Recently this dependence on oil has become less acceptable to Hawaiʻi's residents and their political representatives, especially since the state has the potential to use renewable energy forms.

Hawaiʻi has a good supply of four of the major sources of renewable energy: sunshine, wind, steam produced by molten lava (for geothermal energy production), and burnable crop residue. We even have a few out-of-the-ordinary possibilities for renewable energy production, including ocean thermal energy

conversion, hydrogen power, and perhaps even wave and tidal energy resources.

Hawaiʻi's lawmakers, energy agencies, and citizens are dreaming up and developing ways to bring about change. But we still have much to plan, let alone implement.

In 1962 18% of Hawaiʻi's energy came from burning sugar cane crop residue. This was a very high percentage of energy from a renewable resource even by today's standards. Back then it was stellar. However, as Hawaiʻi's population grew and sugar cane farming decreased throughout the state, this source of renewable energy decreased. By 2000 only 6% of our energy came from renewable sources. At last count, in 2002, it was 5.3%.

The state has made some effort in

Hawai'i's Energy Use Profile

Nonrenewable Sources		Renewable Sources	
89.1%	Petroleum	1.8%	Biomass incineration
5.6%	Coal	1.5%	MSW incineration
		1.3%	Solar water heating
		0.3%	Geothermal
		0.3%	Hydroelectric
		0.03%	Wind
		0.003%	Solar photovoltaic

the past ten or so years to increase the percentage of energy from renewable resources: constructing H-Power, which has been burning some of Hawai'i's municipal solid waste to generate electricity on O'ahu since 1990; developing geothermal power on the Big Island; offering incentives for using solar water heating and buying solar energy panels; and creating wind farms. Crop residue continues to be burned. What happens, however, is that as the population expands, the increased energy demand is met mainly by buying more and more petroleum or coal, so that although more and more energy has come from renewables in recent years, we see little increase in the overall percentage of energy that comes from renewables.

The good news is that the state has set an ambitious, but reachable, goal for itself: In 2004 Governor Lingle signed Senate Bill 2474, known as Act 95, which requires at least 20% of each utility's electricity sales to come from renewable energy sources by 2020. It's the "20/20 Bill" for short. If we can achieve this goal, we will lead the nation. Currently the United States gets only 6% of its energy from renewable energy sources. California already gets 12% of its energy from renewable sources, and 16% of the world's energy comes from renewables. The European Union is shooting for 22% by 2010. Norway gets 99% of its power from

renewable hydropower; New Zealand gets 75% the same way. Denmark banned the burning of coal and now gets 90% of its electricity from wind. Iceland gets 55% of its electricity from geothermal energy. Inspiring examples abound. Challenges remain—cost, systemic inertia to change, working through the rough spots—but there is a negative cost to inaction: continued dependence on a volatile and greenhouse gas–producing energy source with uncertain future availability.

The state has initiated a huge variety of other measures aimed at increasing our energy efficiency. You can read about these on the Department of Business, Economic Development & Tourism Energy, Resources, and Technology Division Web site: www.hawaii.gov/dbedt/info/energy/policy. The same department's Strategic Industries Division also explains many of the state's initiatives.

GLOBAL WARMING

Carbon dioxide is one of the major "greenhouse" gases; acting like a blanket, it helps the Earth trap the energy of the sun. Carbon dioxide is released whenever we break down organic molecules. Living organisms produce carbon dioxide as a waste product from the digestion of meals. Automobiles produce carbon dioxide when they combust gasoline. Burning wood also releases carbon dioxide. The United States, home to almost 5% of the world's population, produces 25% of the world's carbon dioxide emissions.

In 1750 there were about 275 parts per million (ppm) of carbon dioxide in the air. During the last ice age there were 180 ppm. Today there are 381 ppm. Most people finally agree that it's no coincidence that as of 2006, 19 of the 20 hottest years on record occurred in 1980 or later. In a 2006 Time/ABC/Stanford University poll 85% of respondents agree that global warming is happening, and 87% believe that governments should encourage or require the lowering of power plant carbon dioxide emissions.

In 2002 Hawai'i produced 20.4 million tons of carbon dioxide. We produce so much carbon dioxide because we depend on the burning of fossil fuels for 90% of our energy needs. The carbon dioxide we produce may go elsewhere, but carbon dioxide created in other places is blown to us. In the end it all mixes in that common soup we call the atmosphere. Fortunately, in Hawai'i we do have non–carbon-dioxide-emitting energy sources—solar, hydrogen, wind, water, geothermal—and we do have a plan (the Hawai'i Climate Change Action Plan) to use more of them.

Hawai'i is likely to be highly impacted by global warming. When the oceans warm, hurricanes are stronger, sea levels rise, coral reefs die, and ecosystems are correspondingly affected. Hawai'i stands to benefit from becoming an island role model for island Earth in reducing carbon dioxide emissions.

POLLUTION

"Pollution" is anything that has an adverse effect on the health and well-being of humans or ecosystems. When people think of pollution they usually think of poisonous substances, but these are only one of many types of pollutants. Noise and heat can be pollutants. Alien species that arrive with the intentional or unintentional help of humans can become pollutants if they are aggressive and invade native ecosystems. Alien species that cause diseases like avian malaria, dengue fever, West Nile virus, and bird flu can be considered pollutants. Even organisms that occur in a place naturally, like the bacteria in the guts of animals, become pollutants when they become so abundant that they pose a health risk to humans or ecosystems, as is the case in sewage spills.

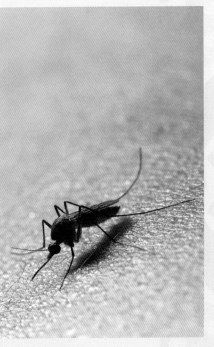

Hawai'i is like other places in having a variety of pollutants acting on very different scales. Smaller-scale pollutants include the coral-smothering invasive alien algae *Gracilaria salicornia*, present only along the south shore of O'ahu—although it is spreading. Also relatively small-scale are the 1,840 confirmed releases of substances from underground storage tanks in Hawai'i between 1987 and 2005, 82% of which have been cleaned up. Other pollutants affect wider areas and more people. When Kīlauea Volcano is erupting on the Big Island and the winds are light, the concentration of sulfur dioxide exceeds federal health standards and poses a health hazard to people all along the Kona coast. On a New Year's Eve in Hawai'i the smoke and particulate matter from fireworks affect people in most of the populated areas of all the islands. On a global scale, Hawai'i's beaches are notorious for collecting marine debris—the

APPENDIX: HAWAI'I'S TEN MAJOR ENVIRONMENTAL ISSUES

floating remains of fishing gear and land-based trash that escaped to sea. Although we cause little of this pollution, we and especially the sea turtles and sea birds that ingest it are adversely affected by it.

Fortunately, Hawaiʻi doesn't have much of the traditionally heavy pollution caused by mining and manufacturing, and it does have trade winds, the steadiest winds on Earth, and ocean currents moving past the islands, both of which dilute whatever air and ocean water pollution we produce.

Unfortunately, however, Hawaiʻi has a history of intensive agriculture involving the use of pesticides and fertilizers. We have cleared native Hawaiian plants from most of the low-lying lands, creating massive erosion. Coastal cities and towns have dense populations and plentiful impervious surfaces: concrete and roads, which generate the most insidious type of pollution to clean up: toxic runoff. Hawaiʻi's ecosystems are always at risk from alien species invasions, and here, where land area is small, an invasion can spell extinction. The next time you wonder whether invasive alien species have adverse effects, just think of the most annoying pests you can: mosquitoes, cockroaches, ants, rats, pigeons, *haole koa*, and *kiawe* are all alien species in Hawaiʻi.

The good news is that government has implemented some programs to curb pollution. An emergency sewage release pipe is being constructed within the Ala Wai so that Waikīkī won't take a direct hit again in an emergency raw sewage dump (only the open ocean will suffer!). The Hawaiʻi State Department of Health's Clean Water Branch monitors coastal and inland water resources and makes the data available to us on a Web site: emdweb.doh.hawaii.gov/CleanWaterBranch/WaterQualityData. The department has also worked with the Coastal Zone Management Program to develop Hawaiʻi's Implementation Plan for Polluted Runoff Control, which was approved by the Environmental Protection Agency (EPA). This led to the creation of the state's Polluted Runoff Control Program. Through this program many community-based projects and groups have conducted programs to clean up and protect our waterways.

The Safe Drinking Water Branch of the State Department of Health tests our drinking water regularly to ensure that it does not exceed maximum contaminant levels for over 90 contaminants. The Wastewater Branch manages our municipal and private wastewater treatment plants. Many organizations, including the National Oceanic and

Atmospheric Administration, have participated in the removal of marine debris. Many nongovernmental organizations also sponsor regular beach cleanups in Hawai'i. Since 1996, 484,000 pounds of marine debris have been removed from the Northwestern Hawaiian Islands.

Now for the bad news. Although many chemical spills have been cleaned up, it is impossible to return a place to the condition it was in before a spill, especially if the place was pristine or wild to begin with. Sewage spills and oil spills continue to occur in Hawai'i. Hawai'i's Hazard Evaluation and Emergency Response Office crews responded to 211 oil spills in 2004. Meanwhile the Polluted Runoff Control Program was not funded in 2005, although the program continued to do what it could using funding leftover from the previous year.

In 2001 the EPA listed 111 Hawaiian waters as "impaired" by some form of pollution. While our drinking water comes mainly from underground sources and surface waters that either test clean or are treated so that they test clean, impaired waters mean impaired ecosystems; whatever is in the stream eventually goes to the ocean and perhaps into our groundwater. In addition, although our drinking water is regularly tested, the test sites are upstream from the user. It is always possible that contaminants get in our water downstream of the test site, as the water travels through street and building pipes.

People have introduced 4,600 species of alien land plants to Hawai'i, of which about 2% have turned out to be invasive. These species have escaped yards, parks, building landscapes, aquariums, pasturelands, suitcases, the soles of shoes, and even bait boxes and spread aggressively in our forests, streams, coastal lands, and coastal ecosystems, challenging native species where they spread.

CULTURE

When a place becomes modernized, the tangible aspects of its culture—language, arts, architecture, foods, medicines—and intangible things—values and beliefs, people's view of their place in the universe, and their attitudes toward the natural world—may be weakened or lost.

Hawaiian culture has not been lost, but modernization has weakened the Hawaiian worldview. The key, perhaps, lies in behaving according to the Hawaiian word *pono*. Among its many meanings are "proper" and "benefit." If we act properly toward nature, nature will benefit us in return.

The core of this book is an appreciation for the value of nature, our place in it, and our dependence on it. Preservation of Hawai'i's natural beauty and resources for your own and future generations should guide your actions as you do 50 simple things to save Hawai'i.

SOURCES

Auditor, State of Hawai'i. 2005. *Hawai'i 2050 Sustainability Task Force Report. A Report to the Governor and the Legislature of Hawai'i.* Honolulu.

Belt Collins Hawaii and Rifer Environmental. 2000. *Hawaii 2000 Plan for Integrated Solid Waste Management.* State Department of Health, Office of Solid Waste Management.

Birkeland, C., and A. M. Friedlander. 2002. *The Importance of Refuges for Reef Fish Replenishment in Hawai'i.* Honolulu: Hawaii Audubon Society Pacific Fisheries Coalition.

Boesch, D. F., R. H. Burroughs, J. E. Baker, et al. 2001. *Marine Pollution in the United States.* Pew Oceans Commission.

Duffy, D. C., and F. Kraus. 2006. Science and the Art of the Solvable in Hawaii's Terrestrial Extinction Crisis. *Environment Hawaii* 16 (11): 3–6.

Eldredge, L. G., and C. M. Smith, eds. 2001. *Guidebook to the Introduced Marine Species in Hawaiian Waters.* Bishop Museum Technical Report 21. Honolulu: Bernice P. Bishop Museum.

Eldredge, L. G., and N. L. Evenhuis. 2003. *Hawai'i's Biodiversity: A Detailed Assessment of the Numbers of Species in the Hawaiian Islands.* Bishop Museum Occasional Papers 76:1–28. Honolulu: Bernice P. Bishop Museum.

Elkington, J. 1998. *Cannibals with Forks: The Triple Bottom Line of 21st Century Business.* Gabriola Island, British Columbia: New Society Publishers.

Energy Information Administration. "Official Energy Statistics of the U.S. Government." www.eia.doe.gov/energy/energybasics101.html.

Environmental Council, State of Hawai'i. *2005 Annual Report.*

Friedlander, A., G. Aeby, and R. Brainard, et al. 2004. Status of Coral Reefs in the Hawaiian Archipelago. Chap. 15 in *Status of the Coral Reefs of the World.* Townsville MC, Australia: Global Coral Reef Monitoring Network.

Hawaii Climate Change Action Plan. 1998. Honolulu: State of Hawai'i Department of Business, Economic Development and Tourism.

Hawaii Energy Strategy: Summary. 2000. Honolulu: State of Hawai'i Department of Business, Economic Development and Tourism, Energy, Resources, and Technology Division.

Indicators of Environmental Quality. 2006. Honolulu: State of Hawai'i Department of Health, Environmental Health Administration.

Kluger, J. 2006. The Tipping Point. *Time*, 3 April.

Luton, C. D., A. M. D. Brasher, and S. L. Goodbred, et al. 2004. The Role of Urbanization and Accompanying Habitat

Alteration in the Establishment of Non-Native Species in Hawaiian Stream Systems. *North American Benthological Society Abstracts.*

Mikulina, Jeff. 2006. It's So Easy: One Bin, Two Bin—Gray Bin, Blue Bin. *Honolulu Star Bulletin,* 22 February.

Miller, G. Tyler. 2005. *Living in the Environment.* Pacific Grove, CA: Brooks/Cole.

Municipal Solid Waste. 2005. Ann Arbor: University of Michigan Center for Sustainable Systems.

Our Changing Climate. 1997. Silver Spring, MD: NOAA Office of Global Programs.

Report to the Twenty-Third Legislature, State of Hawaii. 2006. Honolulu: State of Hawai'i Department of Health, Office of Solid Waste Management.

R. M. Towill Corporation. 1999. *Oahu Municipal Refuse Disposal Alternatives Study: Waste Composition Study.* City and County of Honolulu, Department of Environmental Services.

TenBruggencate, J. 2005. Bush Not Supporting Marine Debris Bill. *Honolulu Advertiser,* Thursday, March 17, 2005.

Thiel, R. E. 1998. Planning and Protecting Coastal Open Space in Hawaii. M. Landsc. Arch. Thesis, University of Washington.

U.S. Environmental Protection Agency. 2003. "Municipal Solid Waste in the United States: 2003 Facts and Figures." www.epa.gov/epaoswer/nonhw/muncpl/pubs/msw05rpt.pdf.

"USFWS Threatened and Endangered Species System (TESS) as of June 20, 2006." ecos.fws.gov/tess_public/StateListing.

Volcanic Air Pollution—A Hazard in Hawaii. 2000. U.S. Geological Survey Fact Sheet, 169–197.

Wilkinson, C., ed. 2004. Executive Summary. *Status of the Coral Reefs of the World: 2004.* Townsville, MC, Australia: Global Coral Reef Monitoring Network.

DIRECTORY OF ENVIRONMENTAL ORGANIZATIONS AND AGENCIES

Department of Environmental Services - Refuse Division
1000 Uluohia St., Ste. 212
Kapolei, HI 96707
info@opala.org

Refuse Division: (808) 692-5358
Recycling Office: (808) 692-5410
Household Hazardous Waste: (808) 692-5411
H-POWER Waste-to-Energy: (808) 682-1359
Waimānalo Gulch Landfill (Nānākuli): (808) 668-2985

Kauaʻi County Recycling Office
4444 Rice St., Ste. 275
Līhuʻe, HI 96766
Phone: (808) 241-6891
Fax: (808) 241-6892
For beverage container deposit program information:
Phone: (808) 241-5112
Fax: (808) 241-6892

Maui Recycling Group
P.O. Box 880852
Pukalani, HI 96788
Phone: (808) 878-6666 (Maui)
Toll Free: (866) 542-2232
Fax: (808) 878-6666
recycle@alohashares.org

Maui Recycling Service
P.O. Box 1267
Wailuku, HI 96793
Phone: (808) 244-0443
Fax: (808) 244-0614
info@mauirecycles.com

Goodwill Industries of Hawaii, Inc.
2610 Kilihau St.
Honolulu, HI 96819-2020
Phone: (808) 836-0313
TTY: (808) 526-3913
Fax: (808) 833-4943
info@higoodwill.org

Hawaiʻi Audubon Society
850 Richards St., Ste. 505
Honolulu, HI 96813-4709
Phone: (808) 528-1432
Fax: (808) 528-1432
hiaudsoc@pixi.com

Hawaii Bicycling League
3442 Waiʻalae Ave., #1
Honolulu, HI 96816
Phone: (808) 735-5756
Fax: (808) 735-7989
bicycle@hbl.org

Hawaiian Ecosystems at Risk Project (HEAR)
P.O. Box 1272
Puʻunene, HI 96784
webmaster@hear.org

Hawaiian Humane Society
2700 Waiʻalae Ave.
Honolulu, HI 96826
Phone: (808) 946-2187
Fax: (808) 955-6034

Na Ala Hele

Big Island:

Irv Kawashima
Na Ala Hele Trails and Access Specialist
Department of Land and Natural Resources
19 East Kāwili St.
Hilo, HI 96720
Phone: (808) 974-4217
ikawashima@dofawha.org

Kaua'i:

Craig Koga
Na Ala Hele Trails and Access Specialist
Department of Land and Natural Resources
3060 'Eiwa St., Rm. 306
Līhu'e, HI 96766-1875
Phone: (808) 274-3442
Fax: (808) 274-3438
craig.c.koga@hawaii.gov

Maui:

Mark Peyton
Trails and Access Technician
Trails Volunteer Coordinator
DLNR-DOFAW-NAH
54 S. High St.
Wailuku, HI 96793
Phone: (808) 873-3509
Fax: (808) 873-3505
mark.a.peyton@hawaii.gov

Maui/Molokai/Lāna'i:

Torrie Nohara
Na Ala Hele Trails and Access Specialist
Department of Land and Natural Resources
54 South High St., Rm. 101
Wailuku, HI 96793
Phone: (808) 873-3508
Fax: (808) 873-3505
Torrie.L.Nohara@hawaii.gov

O'ahu:

Curt A. Cottrell, Na Ala Hele Program Manager
Division of Forestry and Wildlife
Na Ala Hele Trail and Access Program
Department of Land and Natural Resources
1151 Punchbowl St., Rm. 325
Honolulu, HI 96813
Phone: (808) 587-0062
curt@dofaw.net

Aaron Johnson Lowe
Na Ala Hele Trails and Access Specialist
Department of Land and Natural Resources
2135 Makiki Heights Dr.
Honolulu, HI 96822
Phone: (808) 973-9782
Fax: (808) 973-9781
alowe@hawaii.rr.com

The Nature Conservancy Hawai'i
923 Nu'uanu Ave.
Honolulu, HI 96817
Phone: (808) 537-4508
Fax: (808) 545-2019
hawaii@tnc.org
volunteer_hawaii@tnc.org

Recycle Hawai'i
P.O. Box 4847
Hilo, HI 96720
Phone: (808) 329-2886 or
(808) 961-2676
info@recyclehawaii.org

Sierra Club
1040 Richards St., Rm. 306
Honolulu, HI 96813
P.O. Box 2577
Honolulu, HI 96803
Phone: (808) 538-6616

United Cerebral Palsy Association of Hawaii
414 Kūwili St., Ste. 105
Honolulu, HI 96817-5050
Phone: (808) 532-6744
Fax: (808) 532-6747
Toll Free: (800) 606-5654 (Hawai'i only)

Vanpool Hawaii
711 Kapi'olani Blvd, Ste. 985
Honolulu, HI 96813
Phone: (808) 596-VANS (Oahu)
(800) VAN-RIDE (Neighbor Islands)
Fax: (808) 596-2056
vanpool@vpsiinc.com